David Brown

TRACTORS

A BRITISH LEGEND

David Brown

T R A C T O R S

A BRITISH LEGEND

COLIN HOLWELL

Japonica Press

Contents

First Published 2003

ISBN 1 904686 01 X

A catalogue record for this book is available from the British Library

Published by
Japonica Press
Low Green Farm, Hutton, Driffield,
East Yorkshire, YO25 9PX
United Kingdom

Book layout by Banks Design

About the Author

Colin Holwell was born just before World War II, the son of a tenant farmer at Hillcroft, Easton on the Hill, near Stamford, Lincolnshire.

He has been a tractor enthusiast for as long as he can remember, having learnt to drive the Standard Fordson at the age of eight. In 1955 he spent a year at the Northamptonshire Agricultural College at Moulton, Northampton.

His fascination with tractors and machinery led him to invent a bale loader for picking up small round bales made by the Allis Chalmers Roto-Baler – this was patented in 1959.

In 1969 when Colin's father and uncle retired, he was granted the tenancy of Fairview Farm, Easton on the Hill, where the family has farmed since 1932, and where he still farms today.

Colin collects vintage tractors and enjoys participating in Vintage Rallies, where he is a well-known commentator. He is also a popular guest speaker at Vintage Club meetings around the country. He has a vast fund of knowledge on all makes of tractor, but especially David Browns.

Acknowledgements

In writing this book I have received help and encouragement from a lot of people, all of whom I thank very much indeed.

Special thanks must go to my wife Kathleen and daughter Julie for helping me with this project. I sincerely thank my daughter Anne and granddaughter Kayleigh for teaching me my very limited computer skills. Once I have learned to spell, I can apply for a job in a one finger typing pool!

I would also like to thank Michael Hart, who was at one time a David Brown sales representative and has a vast knowledge of the subject; Raymond Townend for allowing me access to his extensive archives; The Rural History Centre, University of Reading, for allowing me to look through their vast photographic history of David Brown Tractors and giving permission for some of their black and white prints to be used in this book; The Tolson Museum, Huddersfield, for information on and photographs of the David Brown valveless car.

Many thanks to Neil Singleton who, together with his two sons, Nigel and James, has provided much technical information.

Also many thanks to Peter D. Simpson for his many excellent photographs.

Help has also been received from Phillip Addy and family, George Chambers, Bob Marsh, George McDonald, Peter Murray, Robert Moorhouse, Ron Petch, John Simpson, Ron Smith, Patrick Thomas, Geoff Warren, Sonny Smith and Geoff Wolfenden. I thank you all.

Foreword

I would like to think that this book, researched and written with great enthusiasm by Colin Holwell, will bring back happy memories to everybody who has, like myself, been involved with David Brown tractors.

My early involvement started with a ride to Newark on a local farmer's Cropmaster tractor, to collect brewers grains. In 1966, I joined Newark Farm Machinery to sell the new range of Selectamatic tractors. I found selling these tractors on their merits of two speed PTO units and a superior hydraulic system an interesting challenge. The most rewarding aspect of tractor sales came with the brown and white tractor – the hydraulic lift did not go down when the engine stopped, as it did on the red and grey tractors.

In David Brown Tractors, Colin has covered every model in a way to suit both the David Brown enthusiast and anyone interested in vintage tractors. I wish him great success with this book.

Michael Hart
Collingham
Notts

The David Brown Story

THE DAVID BROWN STORY starts in 1860 when the first David Brown started making wooden gear patterns for the textile industry. From small beginnings, employing just two people, the workforce increased and soon David Brown's two sons, Ernest and Frank, were employed. Later his younger son, Percy, joined the firm.

Two views of Park Works

PARK WORKS

The main reception at Park Works Huddersfield undergoing repairs and maintenance in April 2002

By the turn of the century the company had grown to such an extent that larger premises were needed. Frank Brown, son of the founder, who was now in charge, had been searching for a suitable location when Park House at Lockwood came onto the market. This property was beside the main Huddersfield to Sheffield railway line, which was an ideal situation. Manufacturing began in the grounds and the site eventually became the famous Park Gear Works.

Having come though World War I with full order books, the company had consolidated and was able to survive the depression of the 1920s when many other firms did not.

After World War I, a foundry was opened on site. David Brown were now able to cast their own gear blanks instead of buying them in. The firm was also quick to realise that the future was going to be in steel gears. David Brown engineers were pioneers in this field.

David Brown junior, Frank's son, joined the firm in 1921 as a 17-year-old apprentice, he rose through the ranks to Foreman, Assistant Works Manager, Director and in 1932 he became Managing Director. In 1934 the old Cammell Laird tyre-rolling mill at Penistone, approximately 12 miles from Lockwood, was purchased. This site, which was also on the Huddersfield to Sheffield railway line, was converted to a steel and bronze foundry.

In the mid 1930s David Brown went on a business trip to America. While he was there, he visited several farms and was particularly impressed by how the American farmers had developed and utilised tractor power. A very high proportion of tractors used in Britain were American imports. David Brown realised that there must be a market for a British built machine and his company, with their experience in building gearboxes and transmissions for the automotive industry, were ideally positioned to build a tractor. He knew that they had a good name and were busy supplying all different products to other firms as components but they did not actually make a finished product themselves.

VALVELESS CAR
Between 1908 and 1915 David Brown produced a valveless car whose rugged hill climbing qualities made it especially popular amongst the pioneer motorists in Africa and Australia

Harry Ferguson

Harry Ferguson was born on the 4 November 1884, on the family farm at Growell, near Dromore, Co Antrim, Northern Ireland. He was the fourth child of James and Mary Ferguson. From his school days, he was not very interested in the drudgery of nineteenth-century farming in Northern Ireland but he liked all things mechanical.

LAKESIDE FARM
Lakeside Farm, Growell,
Dromore, County Down

HARRY FERGUSON
Portrait of Harry Ferguson

He left school at 14 and after a short period on his parents' farm, he joined his brother as an apprentice engineer. He soon became known locally for his talents in motor racing with both cars and motorcycles. By 1908 he had built his own aeroplane, which made him the first British man to build and fly his own plane.

In 1911, he left his brother in order to set up his own small garage in Belfast. It was not long before he outgrew these premises and moved to a larger site at Donegal Square East. He eventually took on the agency for the Overtime tractor. Although the Overtime was reliable and worked well, Ferguson had ideas for better ways of attaching implements to tractors rather than having them pulled by a length of chain and relying on the tractor's weight for traction. He carried out many experiments by using a plough that he had designed coupled to a Ford Model T car with an eros conversion that turned the car into a tractor (later a Fordson Model F tractor was used). He eventually came up with the design of a small tractor that incorporated a linkage which, after further refinement, was to become the world's first three-point linkage to make tractor and implement become one unit. The hydraulic linkage also maintained implement depth and so did away with the need for depth wheels. The implements weight was transferred onto the driving wheels of the tractor and so the need for a heavy tractor was eliminated. Another feature of the linkage was that it prevented the tractor from rearing up if a hidden obstruction was encountered by the implement.

This tractor was painted black hence its name the Black Tractor. Many of the transmission components were made by David Brown at Park Works, Huddersfield, which is where the two great names came together.

Having proved that the system of implement depth control worked and met a criteria that he had set from the outset that 'the tractor should weigh no more than a shire horse but be able to pull twice as much', Harry Ferguson went on to design a tractor, similar to the Black Tractor, which could be mass produced. He had to find a manufacturer to take on the task. Eventually, he came to David Brown who, having made numerous components for the Black Tractor, was interested in building an agricultural tractor.

An agreement was reach whereby David Brown would manufacture the tractor and Ferguson would be responsible for design and sales.

David Brown Ferguson

Production of the Model A started in 1936 at Park Works, in space rented from the parent company. Ferguson also rented an office in Huddersfield so that he could oversee the developments.

C BLOCK
– *the last building to be erected on the Meltham Mills site by the United Thread Company*

The tractor did not sell as well as expected in the first year and friction soon built up between the two tractor pioneers as to the reason why.

In 1937, Harry Ferguson Ltd, who were responsible for sales and design, began to run into financial problems, which led to a new company being formed called Ferguson Brown Ltd. Harry Ferguson and David Brown were joint Managing Directors. By1938 ten tractors per day were being produced.Harry Ferguson went off to America in 1939 to demonstrate his tractor and implements to Henry Ford. The result of this meeting was the now famous golden-handshake agreement whereby Ford would produce a completely new, small Ford tractor incorporating the Ferguson hydraulic system.

This left David Brown free to develop his own tractor. Once again a new company was formed, David Brown Tractors Ltd. This company took over the remaining stocks from Ferguson Brown Ltd.

In the meantime, David Brown, who was having trouble with his father over space for tractor production at the Park Works factory, had acquired the disused thread mills at Meltham Mills.

Meltham Mills is about five miles from Huddersfield. The name originates from a corn mill built in 1760 by a Nathaniel Dyson, who went on to enlarge the premises in 1786. By the early 1800s, Jonas Brook Bros Ltd had established a factory nearby to produce sewing cotton, this also flourished and by 1822 they had erected some large additional buildings. In 1845 Jonas Brook Bros Ltd acquired the original corn mill premises, which were then converted for textile use.

In the depression of the 1930s, Jonas Brook Bros Ltd amalgamated with J & P Coats to form United Thread Mills Ltd. The premises continued under this name until 1939, when David Brown Tractors Ltd purchased the 78 acre site. Meltham Mills is situated in a deeply wooded valley in the foothills of the Pennines. German bombers flew over the site on their way to bomb Manchester during World War II but it was never bombed. Luckily it was one of the very few factories involved in producing aircraft parts that was never hit. After the war there was a period of consolidation at David Brown Tractors.

In 1955 the firm of Harrison, McGregor & Guest Ltd was acquired. This company was founded in 1872 originally to manufacture a wide range of horse-drawn farm equipment and barn machinery. Later the horse-drawn implements gradually gave way to tractor-drawn items such as binders, mowers, swath turners, corn drills and numerous other products, all sold under the Albion name. A range of mowers, ploughs, cultivators, balers and a manure spreader were produced during the late 1960s. Production of these was phased out as the factory space was needed for the manufacture of tractor components. Implements were still made in the form of loaders and diggers for the wide range of David Brown tractors now being produced.

Also in 1955 David Brown Tractors were awarded the Royal Warrant of Appointment to Queen Elizabeth II as manufacturers of agricultural machinery.

On four future occasions David Brown Tractors were granted the Queen's Award to Industry, for export

OPPOSITE

The famous saw-tooth roof of the Meltham Mills factory

achievement: 1966, 1968, 1971 and 1978.

In 1974 the company was awarded a Queen's Award to Industry, for technological achievement. This was the first time that an award such as this has been given to a farm tractor. It was in recognition of the Hydra Shift transmission.

The year 1968 saw the introduction of the safety cab and David Brown Tractors were among the first to provide these, becoming available nearly two years before such cabs became compulsory.

In 1965 a new colour scheme had been introduced: Chocolate Brown for the engine and transmission with Orchid White for the panel work. The next milestone in the history of David Brown Tractors came in 1972 when Tenneco International Inc of Houston, Texas bought the company and made it a subsidiary of JI Case. One year later came another colour change to standardise the colour schemes of the two companies. Orchid White was still used for the panel work but Chocolate Brown was replaced by orange.

Tenneco went on to acquire International Harvester of Doncaster in 1985 and this gradually led to another colour change – to red panel work with black engine and transmission. The change back to red was the final colour change. Meltham Mills ceased tractor production in 1988, after 52 years.

LEFT

Durker Roods Hotel – formerly the home of Sir David Brown

Ferguson Model A

THE FERGUSON TYPE A OR MODEL A was the production model of the Black Tractor, which was the first tractor built by Harry Ferguson to prove his theory of implement weight transfer via the three-point linkage that included draft control or depth control through the top link. The Black Tractor was powered by a Hercules engine supplied by the Hercules Corporation of Canton, Ohio, who also supplied the engines for the first Fordson Model F tractors.

PRODUCTION DETAILS

The Ferguson type A or model A, Ferguson Brown built

Built from: 1936\1939
No. produced: 1350

VAK 1 WHEELS
The Ferguson Model A or Ferguson Brown on VAK 1 wheels

The Ferguson Type A tractor, painted in Battle Ship Grey, was first produced in the spring of 1936 at Park Works, Huddersfield. The last few were made at Meltham Mills. The engine used for the first 500 tractors was a Coventry Climax Type E, which ran on petrol or petrol\TVO (Tractor Vaporising Oil). Harry Ferguson disliked TVO as he maintained that most of the tractor troubles of the day were caused by its misuse. The Coventry Climax engine had a capacity of 2010cc and developed 20 horsepower. The Ferguson Type A had three forward gears and one reverse. It had independent wheel brakes and was the first tractor to be produced with hydraulic lift and converging lift arms. John Deere Models A and B had hydraulics but not converging lift arms. The hydraulic pump on the Type A was driven from the gearbox, which meant the tractor had to move for it to work – an interesting challenge for the inexperienced operator! In 1936 the tractor cost £224, complete with hydraulics. A two-furrow plough, a cultivator, a ridger and an inter-row cultivator were available at £46 each.

By the end of 1937 the engines were of David Brown manufacture; Coventry Climax were concentrating on ministry work and finding it difficult to manufacture small quantities of engines at the same time. The solution to the problem of engine supplies came when David Brown bought the engine casting equipment from Coventry Climax and installed it at the Penistone foundry. The David Brown engine differed little from the Coventry Climax unit: it had a slightly larger sump, a better oil filter and a better air filter.

As the tractor became more widely used, weaknesses started to appear. The gearbox housing was one weakness. David Brown engineers' solution was to change from aluminium to steel, but Harry Ferguson would not hear of this as the use of steel would increase the weight (the weight of the tractor was already 16 $\frac{1}{2}$ cwts). A few cast-iron housings were used as an interim measure until a higher specification alloy casting could be made. All nuts and bolts were of high tensile steel. The majority of the bolts were 7\16 or 5\8 so that the now famous Ferguson spanner fitted all everyday adjustments. Standard rear wheel equipment was ten inch steel wheels or, for row crop work, a six inch steel wheel could be used. Pneumatic tyred wheels were available 4\25x19 fronts and 9x22 rear. From 1938, rear wheel guards (mudguards) were available at extra cost.

ORCHARD VERSION
Ferguson Model A Orchard version

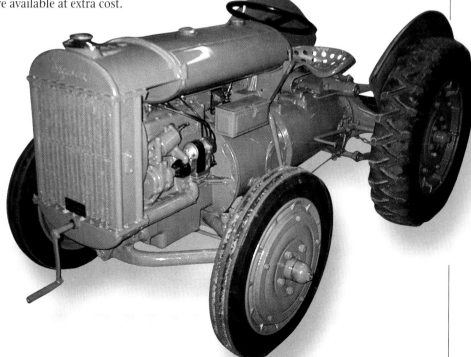

PRODUCTION LINE

*Model A or Ferguson Brown
Production line*

REAR VIEW

*of the Model A showing
Hydraulic lift and PTO
which came over the axle.*

MODEL A DETAIL

*The Ferguson Model A with
side mounted belt pulley and
PTO. Note the foot brake
under the PTO housing, and
the stretch the driver would
have to make to operate the
PTO lever to the front of the
pulley housing.*

VAK 1

THE FIRST TRACTOR to be built by David Brown Tractors that was of David Brown design was the VAK 1 (Vehicle Agricultural Kerosene 1). While building the Ferguson Model A, David Brown was involved in market

research to find out what sort of tractor British farmers wanted. The result of this research was the VAK 1. Introduced at the Royal Show Windsor in 1939, it appeared on the David Brown stand alongside the Ferguson Model A and was advertised as British designed and British built.

VAK 1A MODIFICATIONS
VAK 1 with VAK 1A modifications. Note bullet hole grill and air intake.

BULLET HOLE GRILLE ON VAK 1
These were fitted in the place of the cast grille during the war time austerity period.

VAK 1 Detail

VAK with pulley, PTO and drawbar

VAK 1 Detail

VAK with pulley, drawbar and hydraulics with cable control

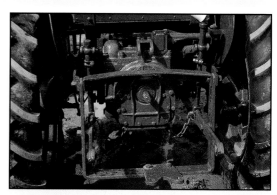

VAK 1 Detail

VAK with PTO, drawbar and hydraulics with mechanical control

PRODUCTION DETAILS

VAK 1
(Vehicle Agricultural Kerosene 1)

Built from: 1939\1945
No. produced: 5300

The new factory at Meltham Mills was coming on stream, everything was set for tractor production and orders for some 3000 tractors were reported to have been taken by the end of the show. Then came World War II and the Meltham Mills factory had to be used for the war effort, which meant that the 3000 tractors ordered at the Royal Show took several years to build – not only was factory space hard to find but labour and materials were also in short supply.

The David Brown VAK 1 was the first of a long line of tractors in the red livery of the famous Hunting Pink. It had a four-cylinder engine which ran on petrol/TVO and developed 35 hp (which would have been quite a powerful tractor for wartime Britain when the main opposition, the Fordson Model N, was 25 hp). The VAK 1 had four forward gears and one reverse. It had independent handbrakes for headland turns. What was described as a streamlined bonnet and windshield (later to be called a scuttle) offered some weather protection for the driver, which had been unheard of in 1939. The wheel track could be adjusted by means of turning the wheel centre dish and using spacers.

Early tractors had a cast radiator grill with horizontal bars. These were expensive to make and very fragile, so later models had what became known as the bullet hole grill, which was sheet steel with round holes cut in and rolled to shape. This was both more durable and cheaper to make during wartime restrictions.

Optional equipment included electric lighting and starting, a rear mounted pulley, PTO with $1\,^3/_8$ inch power take-off shaft and, of course, hydraulic three-point linkage, the lower links of which were parallel as Harry Ferguson still held the patent for converging lower links.

Ten VAK 1s about to be delivered – all on steel wheels, most with pulley and PTO

PRODUCTION
VAK and VTK Production lines

SERVICE FLEET: *David Brown Tractors service fleet, note the war time head light masks*

PROTOTYPE

Prototype VAK 1
Disc Harrowing

The David Brown Agricultural tractor - VAK 1/100

ABRIDGED SPECIFICATION

ENGINE
Four cylinder monobloc with wet cylinder sleeves. Detachable cylinder head carries all valve gear. Bore 3½", Stroke 4", 26 b.h.p. at 1,300 r.p.m. (1st governor position): 35 h.p. at 2,000 r.p.m. (2nd governor position).

CLUTCH AND CONTROLS
Borg & Beck single dry plate. Independent control, either by pedal, or by hand lever operated from seat or from rear of tractor.

GEARBOX AND DIFFERENTIAL UNIT
Self contained and permanently aligned 4 speed and reverse unit incorporating spiral bevels and differential.

REAR AXLE
Rigid casing surrounds the differential and supports, at its outer ends, self-contained final reduction units. Also provides mounting for power lift, power take-off and pulley units.

WHEEL EQUIPMENT
Pneumatics: Front 6.00 - 19. Rear 9.00 - 28. Front 4.50 - 10. Rear 9.00 - 24.
Steel Wheels: Front, 26" rim diameter with fin. Rear, 38" rim diameter, 10" wide. Fitted with twenty 5" steel Spuds. Rowcrop Rear, 40" rim diameter fitted with two-position steel spuds.
Ridge riding front wheels.
Rear wheels with retractable spuds.
Road bands.
Front and rear wheel weights.

DRAWBAR
Exclusive design with 22" lateral adjustment in 1" stages and variable height from 10¾" to 19¾" in 1½" steps. Attachment available to suit Power Lift Unit.

POWER LIFT

Illustrated above. Includes linkage and levelling lever for unit principle implements. Hydraulically operated and controlled by finger tip lever from driving seat. Manual width control device now fitted as standard equipment. Adjustable top link with overload release, optional extras.

ELECTRICAL EQUIPMENT
Optional extras include: Electric starting.
Agricultural lighting - one front and one rear (both head lamp tyre).
Industrial lighting - two head lamps (with pilot bulbs) and tail lamp.

TRACK ADJUSTMENT
Patent common centres giving a combination of 16 positions. Rear wheel track adjustable from 44" to 60" and front wheel track from 46¼" to 57¼".

WEIGHT
With power lift ... 31 cwts. Without power lift ... 28 cwts.

DIMENSIONS

Overall length	105"	Width	66½"
Height	45½"	Wheelbase	64½"
Turning radius (unaided)	8'4"	Ground clearance	16"

POWER TAKE-OFF
Belt pulley:- 8½" diameter x 6½" wide.
Power take-off shaft extension:- Standard spline S.A.E.6B.

VTK and VIG 1/100

Serving on every Continent

WHEN WAR BROKE OUT TOWARDS THE END OF 1939, the ministry realised that military airfields were going to be needed quickly. They approached David Brown Tractors, who were already tooling up for the war effort, to see if they could produce a track-layer tractor suitable for airfield construction that could be used later as an aircraft recovery and towing tractor. The result of this consultation was the VTK (Vehicle Tracked Kerosene) – basically a VAK 1 on tracks with a heavy sub-frame. A lot of the mechanics had to be altered. Eventually a crawler was produced to fit the bill. Some vehicles were fitted with front mounted Hesford winches.

PRODUCTION DETAILS

The VIG 1/100

Built from: 1941\1944
No. produced: 2400

THRESHERMAN CONVERSION

A VIG-100 which was converted to a thresherman in 1974 by Selby Watson of Belmont Street, Oadby, Leicstershire. At some time in its life it had had its front mud guard cut back. It is seen here driving thresher drum and baler.

VIG 100

Note the rear wheel weights and the speedo driven from the front wheel

As the airfields became operational and tractors were used in their aircraft towing and recovery role, they proved to be less than ideal. The first disadvantage was that the steel tracks were making a mess of the aprons and breaking them up when parking planes. The second failing was lack of speed. After a raid, the most damaged aircraft would land first and would need recovering, so out would come the VTK. The runway could be two miles long and if the stricken plane was at the far end, the recovery track-layer, which could do only about four miles per hour, took half an hour to arrive. In the meantime, the rest of the aircraft waiting to land were running out of fuel.

Back to the drawing board! It was decided that a rubber-tyred wheel tractor was the answer, so the VIG 1/100 was born.

The VIG 1/100 (Vehicle Industrial Gasolene 1) started life as a reworked VTK. The tracks were dispensed with and large industrial cast-iron wheels were fitted with tyres, large tyres at the front and 12.75x24 at the rear. The heavy sub-frame was retained to provide extra strength and rigidity. It also provided a frame into which weights could be placed. At the rear was a really heavy winch, of David Brown design and manufacture, that carried 100 feet of $^5/_8$ inch wire rope, which gave the VIG 1/100 an impressive 10,000 lbs of pull. Behind the winch could be fitted either a towing hook or a four height pin hitch. It

had four forward gears and one reverse, 12-volt lighting and starting – a truly heavy recovery tractor with a basic weight of 3.5 tons that could be ballasted up to 7 tons. Once these conversions had been proven, and were up and running, outside engineering firms were engaged to do the remaining conversions, with parts supplied by David Brown. One of the main engineering firms employed in this work was Manns of Saxham, now famous for Claas farm machinery. As each lot of VIG 100 tractors were delivered another lot of VTK crawlers were returned for conversion. The reason for outside firms being employed to do the conversion work was that David Brown Tractors were by now fully committed to the war effort so factory space and labour were at a premium.

What seemed to be the perfect aircraft tug had one more problem to overcome. When fully ballasted it weighed over 7 tons and recovering an aircraft probably weighing 20 tons made a total of 27 tons of machinery to move, unsurprisingly the clutch did not last long. It was back to the drawing board again and the David Brown fluid flywheel was born. This solved the problem.

After the war, a few VIG 100s had front mounted pulleys fitted so that they could be used with threshing sets. They were then called Threshermen – a real case of turning swords into plough shares.

REAR VIEW
VTK rear view showing massive sub chassis with winch rope emerging by the drawbar

FRONT VIEW
VTK front view showing heavy winch and cable route to the rear

VIG 100 CONVOY

BOMB TRAILERS: *A typical war time scene, a VIG 100 with bomb trailers*

A WINTER SCENE: *A VIG 100 doing what it is good at*

OIL TANK: *VIG 100 showing the oil tank to the left side of the dash which kept the fluid fly wheel transmission topped up*

SIDE VIEW: *VTK side view showing track geometry*

DB4

DAVID BROWN'S FIRST ATTEMPT AT PRODUCING A CRAWLER TRACTOR was the VTK which could hardly be described as successful. The second attempt was the DB 4. This tractor was built, on behalf of the British Government, under licence from drawings supplied by the Caterpillar company of America. It was very similar to the Caterpillar D 4.

DB4 *stored away awaiting restoration*

DB4
Publicity shot of the DB4 with dozer blade

PRODUCTION DETAILS

The David Brown DB4

Built from: 1942\1949
No. produced: 110

The DB 4 was the first tractor that David Brown built that used a diesel engine. The engine, which developed 38.5 hp, was supplied by Dorman. The tractor had a five speed gearbox with clutch and brake steering. These tractors, many of which were fitted with bulldozer blades, were used by the British army engineers for construction work. They made a name for themselves for their rugged reliability, especially on the Normandy beaches and in the North African campaign of World War II.

VIEW *from the drivers seat of a DB4*

DB4 *at work*

VAK 1A

AS MENTIONED PREVIOUSLY, DUE TO THE WAR EFFORT, labour, materials and factory space were very limited during production of the VAK 1. However, during this time valuable experience had been gained in producing aircraft towing tractors and industrial crawlers, both of which we shall discuss later. In 1945 the war was at last over, which meant that the wealth of experience gained from the war effort could be used to improve on the VAK 1. The VAK 1A was developed.

PRODUCTION DETAILS

The David Brown VAK 1A

Built from: 1945\1947
No. produced: 3500

AIR-INTAKE
Modified air-intake on VAK 1A, which came from the manifold (The VAK 1 had a mushroom dome on the air-intake.)

The main improvements included a much better engine lubrication system; a more precise governor for better control of engine speed; and an automatic load controlled 'hot spot' for a quicker engine warm-up so that Kerosene (TVO) could be used more efficiently. The mushroom on the air-intake disappeared. The engine air-intake now came from the exhaust manifold so that once the engine was running it drew warm air, which also helped to vaporise the TVO. A turnbuckle top link, a David Brown patent, was introduced at this time. Today, this link is used by most tractor manufacturers.

Several VAK 1s were upgraded with many of the VAK 1A's improvements. The easiest way to identify the difference is from the axle – the VAK 1 has a round tubular front axle whereas the VAK 1A has what is known as a square front axle. The 1A also has a pressed steel radiator grill with horizontal bars. A horizontal brass badge was displayed just above the grill.

VAK 1A
VAK 1A front view, showing pressed steel grill with horizontal brass badge above

Cropmaster VAK 1C

Every tractor manufacturer seems to have one model which made their name and put them on the map, so to speak. In David Brown's case the Cropmaster was that tractor. The Cropmaster was the result of 10 years' experience of building agricultural and industrial tractors for both the military and all sorts of industries that needed a variety of towing tractors, from heavy 7 ton plus tractors to light runabouts.

REAR VIEW
Rear view of the Cropmaster showing square frame drawbar, two speed pulley, PTO, and the hand clutch between the left mudguard and the seat

EARLY CROPMASTER
An early Cropmaster with round front and name painted on the bonnet side (Note that there are no electrics.)

CROPMASTER AND BAMFORD FY4 SPREADER
An annual chore on a dairy farm: muck spreading with a diesel Cropmaster and Bamford FY4 Spreader

PRODUCTION DETAILS

The David Brown Cropmaster

Built from: 1947\1953
No. produced: 59,800

Introduced in 1947, the David Brown Cropmaster, or David Brown VAK 1C, heralded a new policy. Instead of selling a basic tractor and adding on all the extras such as power take-off, pulley, hydraulics and electrics, all would be included in a basic or standard price.

In appearance the Cropmaster looked more balanced and more business-like than its predecessors, with many novel features. The name badge was vertically positioned down the centre of the radiator grill. The name David Brown Cropmaster was painted on each side of the bonnet (this was replaced by a chrome badge on later models). The tractor had a two-piece mainframe as opposed to the single one on most of the VAK 1 As. Under the bonnet there was a four-cylinder engine with a $3^1/_2$ inch bore by 4 inch stroke of 2523cc. The whole of the engine and transmission is offset two inches to the left in order to put slightly more weight on the land wheel when ploughing. As any ploughman will tell you, it's not a problem to get traction with the furrow wheel because the tractor naturally slopes towards the furrow, which increases the weight on that wheel. Moving further towards the rear of the machine, there are much larger mudguards over 28 inch rear wheels. A choice of four speed or six speed gearboxes with two speed power take-off and a two speed pulley as standard equipment – two speed power take-off and pulley would have been unheard of in 1947. The drawbar is now a sturdy pin on a fair interchangeable with the lift arms so that the tractor can easily be

FULL LIGHTING
An early Cropmaster with full lighting – the near-side headlight would have been an extra

changed from drawbar work to mounted implement work. The lift arms could be locked in the 'up' position and the 'halfway down' position by means of large stable-door type bolts. There was a double seat upholstered in waterproof material so that the driver's assistant could ride in safety and comfort – a safety feature not seen on many tractors of the period. To the left or near-side of the bench seat was a hand clutch which could be used on level ground when hitching trailed implements. The driver dismounted, and from the back of the tractor operated the hand clutch with his left hand. Once the gear lever had been placed in reverse position, this freed his right hand to hold up the implement drawbar for the tractor, now controlled by the left hand, to be reversed. In practice the hand clutch was also used when the driver was picking up bales, sheaves, or anything from the nearside. Instead of walking round to the offside, he would stand on the drawbar and operate the tractor from the back. What would the Health and Safety Executive say today?

Although the Cropmaster was sold as the most comprehensively equipped tractor, with four sources of power as standard, drawbar, hydraulics, power take-off and pulley, there were two prices. In 1949 the Cropmaster six speed had a list price of £475 with 6-volt electric starting and lighting. The price without electric starting and lighting was £453. The same price list quoted the Cropmaster as weighing 29.5 cwts. It had a hydraulic lift capacity of 2700 lbs, a ground clearance of 16 inches and a turning radius of 10 feet 6 inches.

In 1949 a diesel engine Cropmaster was introduced,

which developed 34 hp. David Brown Tractors were the first major tractor manufacturers to make their own diesel engine in-house. The diesel engine was interchangeable with the petrol/TVO engine. Many diesel engines were sold as retro fits. The Cropmaster diesel had 12-volt electrics and was readily identified by having two large 6-volt batteries mounted on the chassis, one each side, and mounted higher than on the TVO model, which had only one 6-volt battery. The diesel tractor also had two headlights as standard; the second headlight on the TVO model was an extra.

When the diesel Cropmaster was introduced, the front bolster casting was changed from a round front to a square type. The gear driven dynamo was replaced by a fan belt driven 12-volt one. Two headlights became standard on all Cropmasters.

VINEYARD CROPMASTER

David Brown Tractors produced a vineyard model Cropmaster from 1947 to 1953. Most of these were exported but a few stayed in the UK. Although they looked very narrow, in fact they were only seven inches narrower in overall width on the narrowest wheel settings. A normal tractor is 61 inches wide; the vineyard model was 54 inches wide, which meant that the minimum rear track was reduced from 48 inches to 44 inches. The front track was reduced from 49 inches to 39 inches. The Vineyard Cropmaster had much narrower rear mudguards, similar to the VAK 1, and much narrower wheels and tyres.

SUPER CROPMASTER

This was a more powerful Cropmaster. It shares all the features of the standard Cropmaster plus the following, starting at the front: chaff screens on the radiator grill, two headlights as standard – this tractor had 12-volt electrics whereas the ordinary Cropmaster had only 6 volts. Moving slightly further back: the Cropmaster had a mushroom dome on the air-intake, a rain trap on the exhaust pipe and full bonnet sides which were vertically fluted. Under the bonnet, the engine had a $3\,^5/_8$ inch bore by 4 inch stroke and ran at higher revs, which gave it three more brake horsepower. The rear mudguards were slightly larger to accommodate the 13x28 tyres and wheels. When combined with the increase in engine speed, this made all the gears that bit faster than on the basic Cropmaster.

Altogether, nearly 5000 Super Cropmasters were produced between 1950 and 1952.

PRAIRIE CROPMASTER

The Prairie Cropmaster was similar to the Super Cropmaster. The Prairie Cropmaster had 12-volt electrics with full bonnet sides, large wheels and tyres, two headlights but no scuttle. Instead of the dual seat, a single pan seat was fitted. The driver sat astride the transmission with a clutch on the left and independent foot brakes on the right. Introduced in 1951, the Prairie Cropmaster, as the name implies, was aimed at the North American and Canadian markets. The diesel version was very successful. It was produced between 1951 and 1954.

VINEYARD CROPMASTER
Vineyard Cropmaster with under slung exhaust (This tractor is seven inches narrower than the normal Cropmaster.)

CROPMASTER TVO

David Brown Cropmaster TVO

CROPMASTER
Cropmaster with front mud-guards

ORCHARD MODEL
The Orchard Model Cropmaster

Taskmaster VIG 1AR and VID 1AR

BEFORE THE TASKMASTER, the light industrial tug market was catered for by upgrading an agricultural tractor and adding larger or better brakes and better lighting.

VIG 100 1C
THRESHERMAN
(Note the pressed steel wheels, which were a post-war alternative as wartime models had cast wheels.)

REAR VIEW
Another view of the Taskmaster at left of page, showing heavy winch and ground anchor

MOD LIVERY
Taskmaster in MOD livery with passenger grab handles on each mudguard

PRODUCTION DETAILS

**David Brown Taskmaster
VIG 1 and VID 1**

Built from: 1948\1965
No. produced: 2752

In 1948 the Taskmaster, a specialist industrial tractor intended for light haulage, was introduced. It incorporated the Cropmaster's engine, gearbox (four speed on the early models and six speed on the later ones) and rear axle. The original inboard brakes were used for parking. Additional outboard brakes operated by a single foot pedal were used for the main brakes. The steering was also stronger and heavier than on its agricultural counterpart and there were heavy mudguards both back and front. Many Taskmasters were fitted with a fluid drive torque converter in the transmission. Towards the end of the production run, the 950 tractor was used as the skid unit.

VIG 1C

This tractor was developed as a wartime Air Ministry tractor. It incorporated the Cropmaster engine with a fluid drive torque converter in the transmission. Basically it was a much heavier Taskmaster. Underneath was the heavy chassis of the wartime version and at the back was a large winch complete with 100 feet of 5/8 wire rope and a massive sprag or ground anchor. Some of these tractors had front mounted belt pulleys, fitted and, as their predecessors, they were known as Threshermen.

The David Brown VIG 1C was built between 1952 and 1958. In total, 330 were produced.

BALLAST WEIGHTS
Another view of the same Taskmaster, showing the rear ballast weights (The hydraulics are operated by a crankshaft-driven pump.)

LOADING SHOVEL
Taskmaster fitted with loading shovel

DIESEL TUG

Taskmaster Tug with cab – a diesel version

TUG

Taskmaster Tug with cab – could be ex-WD

ON THE DOCKSIDE
Taskmaster moving railway wagons on the dockside

MOVING COAL
Taskmaster Shunter moving coal wagons

HARD AT WORK
A Thresherman hard at work (Note the neatly trimmed thatch on the stack behind.)

HARD AT WORK: *A Super Taskmaster towing a Vulcan at the Farnborough Air Show*

LEFT: AIRCRAFT
RECOVERY TRACTOR
*VIG 1AR – a later version of
the wartime heavy aircraft
recovery tractor, now fitted
with a Cropmaster engine*

RIGHT: DOUGLAS
TURBO TASKMASTER
*(Douglas used concrete bal-
last whereas David Brown
Tractors used steel.)*

CAST WHEELS
*A VIG 1AR, showing cast
wheels and heavy duty winch*

25, 25D, 30C and 30D

BY 1952 THE CROPMASTER, although selling well, was becoming a little expensive compared to the opposition. One reason was because it was so well equipped with many features as standard that were extras on other makes. One manufacturer in the 1950s even quoted a price less lubricating oils.

25D
Two views of a 25D

FUEL TAP
Fuel tap on the 25 (As with the Cropmaster, the tap could be reached from the driver's seat.)

A DAVID BROWN 25
(Note the smaller wheels compared with the 30.)

PRODUCTION DETAILS

David Brown 25, 25D

Built from: 1953\1958
No. produced: 24,742

David Brown 30C, 30D

Built from: 1953\1958
No. produced: 16,073

In the 1940s farmers were desperate for tractors, they had managed through the depressed 1930s, when they could not afford a new tractor. They had managed through the war, when they could not have a new tractor. After the war it wasn't so much the cost but "when can I have it?" Consequently by the 1950s any farmer who wanted a new tractor had got one. The second-hand market was good because a lot of small farmers were looking for good second-hand tractors, which in turn kept new tractor sales buoyant, although very competitive.

To meet the competition, David Brown introduced the 25, 25D and the 30C, 30D range. As these models were cheaper than Cropmasters, the factory could continue production of the cheaper tractor with very little alteration to the production facilities.

The new tractor range carried on with the six speed gearbox, the two speed pulley and the two speed power take-off of the Cropmaster, in fact, early 30C, 30D models also used the Cropmaster dual seat and wide mudguards. The 25, 25D and later models of 30C, 30D lost the dual seat, together with the famous David Brown

DRIVER'S SEAT
View from the driver's seat of a 25

AIR-INTAKE
A TVO Model 25, showing its Cropmaster ancestry with air-intake coming from the exhaust manifold

HARD AT WORK

A 25D hard at work with a rotovator (note later type rear wheel centres)

A LATER 30D
– with larger tyres (The front weights can just be seen under the front casting.)

REAR VIEW
Rear view of 30 series tractor, showing two speed PTO and pulley

scuttle and wide mudguards, in favour of a centrally mounted pan seat and narrow mudguards. The latter became known as shell mudguards.

The 25 was the petrol/TVO model that had a derated Cropmaster engine, which now developed 31 hp. The 25D was the diesel model, which had a derated Cropmaster diesel engine. That also developed 31 hp. Standard tyre equipment was 400x19 front and 10x28 rear. The prices in 1957 were £470 15s 0d (£470.75p) for the 25 and £558 for the 25D.

The 30C had a Super Cropmaster petrol/TVO engine which developed 37.5 hp while the 30D had a Cropmaster diesel engine developing 34 hp. Standard tyre sizes were 600x19 front and 11x28 rear. In 1957 the 30D cost £635.

In 1954, all four models were available with TCU (Traction Control Unit), a device that transferred weight from the mounted implement to the rear wheels of the tractor. The implement still had to have a depth wheel but, by simply turning a large tap, more or less weight could be transferred, according to the requirements of

the conditions. One year later, in 1955, a special hitch was available so that TCU could be used with four-wheel trailers and heavy trailed implements. TCU is sometimes referred to as Two Lever Hydraulics.

The 25, 25D models, produced from 1953 to 1958, were small enough for the large farmer to use for general haulage and runabout duties but at the same time were large enough and powerful enough for the one tractor small farmer. In five years of production 24,742 were made. The 30C, 30D models were produced over the same period of time but not in quite such large numbers: 16,073 were built.

A modified 25 tractor that was five inches narrower and four inches lower than the standard tractor was available. This smaller tractor was to cater for the demand from fruit farmers for a tractor suitable for orchard work. The modification was carried out by Messrs Drake & Fletcher Ltd of Maidstone, Kent, who were main David Brown distributors and dealers, in conjunction with David Brown Tractors Ltd.

The price of the modified tractor in 1953 was £520.

30C, 30D
An early 30C nearest the camera, with an early 30D furthest away (Early 30 series tractors used Cropmaster tin-work.)

AN EARLY 30D
*An early 30D, showing the battery
box mounted much higher than on
the TVO model*

ORCHARD MODEL
An Orchard Model 25 – converted by Drake and Fletcher Ltd of Maidstone, Kent

BROCHURE
David Brown 25 brochure c. 1954

REAR WHEEL DETAIL
A 30D with the later type of rear wheel centre

David Brown Trackmaster 30T, 30TD, 30ITD, 50TD and 50ITD

TRACKMASTER was the name given to the David Brown crawlers, as Cropmaster was to the David Brown wheel tractors.

BADGE: *Trackmaster badge*

LEFT: TRACKMASTER TVO: *Shares the same engine as the Cropmaster*

Trackmaster Brochure

30TD - *The 30TD (This is very similar in appearance to the 50TD but has the smaller four-cylinder engine.)*

DAVID BROWN TRACKMASTER 30T, 30TD, AND 30ITD

The Trackmaster shared the same engines as the Cropmaster – both petrol\TVO and diesel versions were employed. A new six speed gearbox was used, coupled to a stronger transmission, which incorporated controlled differential steering.

When, in 1953, the Cropmaster name was dropped in favour of the 25 and 30 so too was the Trackmaster name in favour of the 30T which was the petrol/TVO version. The 30TD was the diesel version and the 30ITD was the industrial diesel version. This industrial model had an extra pair of track rollers.

30T TVO MODEL - *A 30T – the TVO model of the 30 series crawler*

ENGINE DETAIL
30TD with side panel removed, showing diesel engine with heavy duty starter

RIGHTT: 30 T REAR
Rear end of the 30T, showing the massive drawbar and PTO

50TD with loading shovel, on site clearance

INSET: 50 TD, *using the same engine as the 50D wheel tractor*

DAVID BROWN TRACKMASTER 50, 50TD, AND 50ITD

The Trackmaster 50, as the name suggests, was fitted with a six-cylinder 50 hp diesel engine. This model was not available with a petrol/TVO engine. In 1953, it too lost the name Trackmaster and became either a 50TD or a 50ITD. As with the 30ITD, the 50ITD had an extra pair of track rollers for industrial use. Later models had a larger diameter clutch with redesigned running gear to cope with the power of the larger engine.

30TD WITH HEAVY DUTY RADIATOR GUARD
(These guards were usually found on ITD models.)

50TD

Rear end of 50TD (Note the massive drawbar and PTO.)

TRACKMASTER NARROW

Trackmaster Narrow – TVO model

David Brown 40TD and 50D

THE DAVID BROWN 40TD was the last of the David Brown track-layers. It employed the engine from the 950 wheel tractor mated to the gearbox and transmission from the 50TD. • This combination of components, which were already being produced, made a very reliable and economical 40 hp crawler.

ABOVE
40TD Industrial (This is the last of the David Brown crawlers to use the engine from the 950 tractor.)

LEFT
40TD cultivating, using a Ransome drag

PRODUCTION DETAILS

The David Brown 50D

Built from: 1953\1959
No. produced: 1260

BOTTOM RIGHT:
A 50D name plate

BELOW:
Rear end of the 50D, showing the massive drawbar, four speed PTO and original fire extinguisher

The David Brown 50D or VAD 6 (Vehicle Agricultural Diesel six-cylinder) was the first David Brown tractor to be offered in diesel form only. It was a massive six-cylinder, 50 hp towing tractor which was exported to South America, South Africa and Australia, in particular. A few went to Europe and a few stayed in England.

The 50D model used the six-cylinder engine from the David Brown Trackmaster 50 crawler, which was introduced a year earlier in 1952, coupled to a massive purpose built rear end, designed from scratch for the 50D. The result was a large tractor that the British farmer was not yet ready for. However, Marshall's of Gainsborough were building their MP6, which at that time was the most powerful tractor in production, and selling it overseas for sugar plantation work. The MP6 is the tractor with which the 50D was designed to compete.

The 50D started as the Cropmaster 50 but the Cropmaster name was soon dropped. The 50D was a 50 hp tractor with side mounted belt pulley. Once again, this was unusual as most David Brown tractors had rear mounted belt pulleys. The belt pulley was two speed, the headlights were mounted high on long brackets to keep them out of the way of the belt when the tractor was on pulley work. When a pulley was fitted, the air cleaner was also raised, for the same reason. By 1953, other David Brown Tractors had lost their dual seats but the 50D retained its bench seat. It had no hydraulics at all. What a tractor it would have been if it had. Another unusual feature of the 50D was that it had a four speed PTO. A sliding rear axle was fitted to allow the rear wheel to be moved out for in-furrow ploughing so that the inside wall of both front and rear tyres were in line.

The price in 1957 was £967 10s (£967.50p)

RIGHT SIDE: *50D right side, showing full side panel*

50D: *A 50D showing raised air cleaner to accommodate the belt pulley*

David Brown 2D

INTRODUCED AT THE SMITHFIELD SHOW IN 1955, the 2D was supposed to be all things to all farmers. Its product code was VAD 12 (Vehicle Agricultural Diesel, one/two-cylinder).

FINGER BAR MOWER
2D with finger bar mower

CULTIVATOR: *2D equipped with an inter-row cultivator*

ERADICATOR BARS
Wheel mark eradicator bars to which various tines could be attached (Note the stop control on the bottom left of the radiator screen. This was just right for small boys to pull when the driver wasn't looking!)

2D Industrial (Only a very small number of these were made)

PRODUCTION DETAILS

The David Brown 2D

Built from: 1956\1961
No. produced: 1350

The David Brown 2D, with its engine overhanging at the back, was a very unconventional tractor for the time. The farming press called it the tractor of the future. The factory personnel called it the back-to-front tractor and, no doubt, other people had other names for it.

The idea was to have a light row crop tractor which could do everything. With the engine behind and the driver's seat, just in front of the rear axle but behind any implement which was centrally mounted between front and rear axles, so that the driver had an unimpeded view of the task in hand. With the multitude of implements available it was a truly universal tractor.

It was ideal for the large farmer who required a specialised tool carrier for row crop work in sugar beet and potatoes. Also it was ideal for the small grower or market gardener who wanted a small tractor to do everything, especially those who were thinking of replacing Dobbin. Here was a tractor that was very cheap to run but also very manoeuvrable.

The 2D was a motorised tool bar or tool carrier with an air-cooled two-cylinder diesel engine which was designed by David Brown especially for the 2D. It had the same 3 1/2 inch bore by 4 inch stroke as the Cropmaster tractor. Many parts were interchangeable with the Cropmaster. The 2D had 77 cubic inch capacity and rated at 14 hp. The two aluminium pistons went up and down together and to balance this it had a cast-iron piston in the sump that produced a very smooth running engine. Later the engine was sold as a skid unit for other applications.

The power for lifting the attachments was created by compressed air – the main tubular frame of the tractor being used as a reservoir. Air was supplied by a small compressor mounted on the front of the gearbox. The 2D, with the aid of a tyre inflator, was one of the few tractors of the 1950s that could blow up its own tyres. It had four forward and one reverse gear. Starting was by hand inertia, or electricity at an extra cost of £10. The wheel track width could be adjusted in two inch steps from 48 inches to 64 inches. A rear power take-off, which was also air operated, was available at extra cost.

The basic price of the 2D when introduced was £389 or £399 with electric starting. There were a number of packs available with the tractor and various implements to suit different needs. For instance, the commercial growers pack included a tractor, a tool bar with depth wheels, rear mark removers, ridgers, markers, gang hoes and a plough, all for the price of £514. There were packs for dairy farmers, a row crop pack, a general purpose pack and a market garden pack.

Variations on the 2D theme included a narrow or vineyard model which had an overall width of one metre. Most of the narrower models were for export but, like all export models, a few escaped and were used in this country.

An industrial variation of the 2D was of a shortened length, with weights added at the front to give added stability. Primarily, it was a light towing tractor. At least one was used on the canals for towing boats. Very few industrial versions were made.

2D with reversible plough, showing the wheel settings

Rear view of Vineyard Model 2D with an overall width of one metre

BELOW RIGHT: *2D equipped with a bean inter-row hoe*

FARM MECHANIZATION

The David Brown 2D Tractor

Specially drawn by a
" Farm Mechanization " artist

A report on this new tractor is published on the two
preceding pages.

Above: The
inertia starter,
with control
lever.

Above: The mower, which has
been specially designed for
the tractor, is hitched under the
rear axle and driven from
the p.t.-o. shaft, the coupling
to which is shown at reference
15A. Note that this shaft (reference 15 on main drawing)
is an extension of the shaft
which drives the rear axles.

Below: This plough is the left-
hand one of a pair supplied for
one-way ploughing. It is
attached to the left-hand side
of the tractor backbone and
there are three adjustments,
as shown by the arrows. Gen-
eral-purpose or digger bodies
are available.

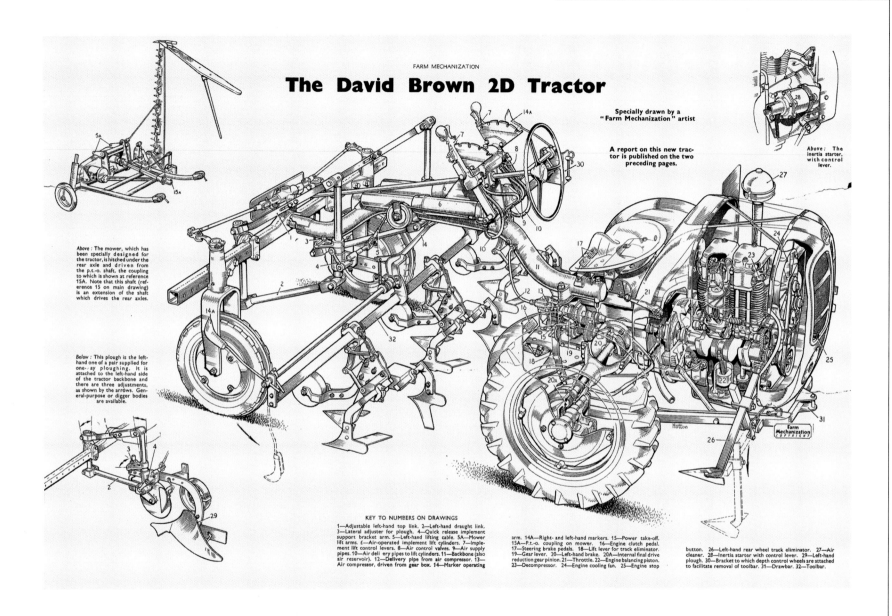

KEY TO NUMBERS ON DRAWINGS
1—Adjustable left-hand top link. 2—Left-hand draught link.
3—Lateral adjuster for plough. 4—Quick release implement
support bracket arm. 5—Left-hand lifting cable. 5A—Mower
lift arms. 6—Air-operated implement lift cylinders. 7—Implement lift control levers. 8—Air control valves. 9—Air supply
pipes. 10—Air deli ery pipes to lift cylinders. 11—Backbone (also
air reservoir). 12—Delivery pipe from air compressor. 13—
Air compressor, driven from gear box. 14—Marker operating
arm. 14A—Right- and left-hand markers. 15—Power take-off.
15A—P.t.-o. coupling on mower. 16—Engine clutch pedal.
17—Steering brake pedals. 18—Lift lever for track eliminator.
19—Gear lever. 20—Left-hand brake. 20A—Internal final drive
reduction gear pinion. 21—Throttle. 22—Engine balancing piston.
23—Decompressor. 24—Engine cooling fan. 25—Engine stop
button. 26—Left-hand rear wheel track eliminator. 27—Air
cleaner. 28—Inertia starter with control lever. 29—Left-hand
plough. 30—Bracket to which depth control wheels are attached
to facilitate removal of toolbar. 31—Drawbar. 32—Toolbar.

BROCHURE:

Spread from a David Brown brochure

David Brown 900

BY 1955, the David Brown 25 and 30 range, being Cropmaster derivatives, were getting a bit dated. The Fordson Major was selling well and something was needed to compete with it. Massey Ferguson were also rumoured to be bringing out a bigger tractor.

DB 900 *(Note the DB badge on the radiator grill is vertical on the non-live drive 900 and the steering wheel is offset. This model was the first departure from Hunting Pink.)*

CUT-AWAY: *900 live drive (This tractor was prepared for exhibition purposes by students at Meltham Mills.)*

PRODUCTION DETAILS

The David Bown 900

Built from: 1956\1958
No. produced: 13,779

FRONT VIEW

Front view of 900 live drive, showing central driving position and horizontal David Brown badge above the radiator grill

David Brown introduced the 900 in 1956. It was a more powerful looking tractor with four engine options: a 40 hp diesel; a 37 hp petrol\TVO; a 40 hp petrol only; and a 45 hp high compression petrol engine. In the past David Brown tractors had been under rated by potential customers who thought the tractor designation was the actual horsepower, for example, that the David Brown 25 was 25 hp when, in fact, it was 30 hp. By calling the latest tractor the 900, the reason for this misunderstanding was eliminated.

This tractor also brought about a new colour scheme, the main tractor retained the Hunting Pink red but the wheels and radiator grills were blue. The front axle was a three section forged one, which looked immensely strong. The outer two sections could be adjusted in or out in two inch stages to adjust the wheel track width. At wider wheel settings kick-back on the steering wheel, which had always been a problem with previous models, was greatly reduced.

The David Brown 900 was the first David Brown tractor to have a distributor type diesel injection pump made by CAV. This pump was the cause of some concern as the failure rate in the first months was very high. The pumps would run on test but would fail after a short time in service. The combined forces of CAV and David Brown engineers could not solve the problem until, quite by accident, it was realised that some of the crates of pumps that were arriving at Huddersfield railway station by passenger train (so desperate were David Brown to get injection pumps) were being dropped on the platform. This set alarm bells ringing. After further analysis of this rough handling it was found that the stresses built up by the machining of the components were being relieved when the pumps were dropped or worked for a short period causing the pumps to seize. The good thing was that the seizure only caused a drive key to shear and no internal damage was done to the engine. The bad thing was that David Brown's reputation took a battering.

The 900 also featured the now familiar David Brown two speed power take-off and pulley, the bonnet badge was vertical down the centre of the radiator grill and, as on previous models, the steering column was offset.

In 1957, live drive, which gave the 900 live power take-off and live hydraulics, was introduced. This was the first David Brown tractor to have a dual clutch; at the same time the steering column was moved to the centre. The main engine frame was changed from the open type (where everybody stored tools and rags) to an enclosed type – a design that the company continued with until the end of tractor production. The front badge was redesigned to be horizontal, above the radiator grill.

DAVID BROWN 903

The David Brown 903 was a three-wheeled high-clearance version of the 900. It had 11x38 rear wheels, instead of 11x28 as on the standard tractor, and a single front wheel set further forward so that under-slung tool-bars could be accommodated. The rear wheel width adjustment ranged from 52 inches to 80 inches. The under clearance of this tractor was a staggering 29 inches. Only a handful of these tractors were built. They were designed mainly for the American market.

ABOVE: *An early David Brown 900 with Sir David Brown seated and his son standing by*
OPPOSITE: *The David Brown 900*

David Brown 950 Implematic, 850 Implematic, 880 Implematic

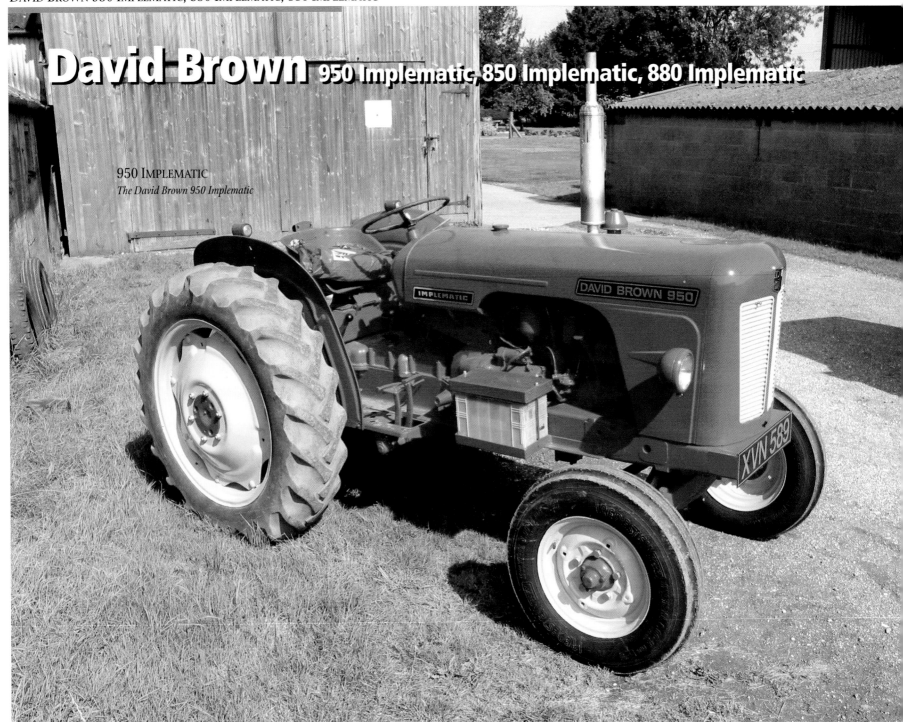

950 IMPLEMATIC

The David Brown 950 Implematic

PRODUCTION DETAILS

David Brown 950 Implematic

Built from: 1959\1962
No. produced: 18,125

David Brown 850 Implematic

Built from: 1960\1965
No. produced: 14,242

David Brown 880 Implematic

Built from: 1961\1965
No. produced: 19,207

DAVID BROWN 950 IMPLEMATIC

This tractor was a David Brown 950 tractor with the ability to handle implements with or without depth wheels. For operating implements with depth wheels, the patented David Brown Traction Control Unit would be used, as with previous models, when extra traction was required. For implements with no depth wheels, the tractors top link operated a Bowden cable, which, in turn, operated an hydraulic valve to maintain implement depth.

In 1961, the 950 Implematic was again improved by having greater front axle clearance and a multi-speed power take-off, providing both 540 and 1000 rpm.

DAVID BROWN 850 IMPLEMATIC

This tractor was available in petrol or diesel form to start with but later only diesel versions were available. The engines both developed 35 hp. This was the last David Brown tractor to have the original engine dimensions of 3 $\frac{1}{2}$ inch bore, with 4 inch stroke. The Implematic hydraulic system, similar to the one employed in the 950, was used: this gave the small tractor an outstanding performance when using mounted implements. The later series with diesel only engines also featured multi-

REAR VIEW - *880 Implematic*

speed power take-off giving 540/1000 rpm; an improved front axle clearance; and, from 1963, height control was included in the hydraulic system. Standard wheel equipment was 11x28 at the rear and 16 inch at the front.

DAVID BROWN 880 IMPLEMATIC

The David Brown 880 started life as an 850 tractor with a 950 engine and was aimed primarily at the one tractor farmer. It developed 42.5 hp from its four-cylinder engine, which gave this small tractor enough power to operate heavy power take-off machinery, which would normally require a much larger tractor. Rotorvators, rotorspreaders, balers and combines all came within its capabilities.

In September 1964, a new three-cylinder engine was introduced – the AD3/40. This engine had a bore and stroke of 3 13/16x4 $\frac{1}{2}$ inches, it also featured a cross-flow cylinder head. The high torque characteristics of the new engine gave it fantastic lugging power. The 880 with its new power unit now replaced the four-cylinder 850 and the 950 Implematic tractors.

850 IMPLEMATIC
The 850 Implematic can handle implements with or without depth wheels

950 IMPLEMATIC
950 Implematic with Hay Bob

ABOVE: **880 IMPLEMATIC**
Early versions were an 850 tractor with a 950 engine

REAR VIEW
Rear view of 850 Implematic

950 IMPLEMATIC

*The David Brown 950
Implematic*

950 IMPLEMATIC

The 950 Implematic can handle implements with or without depth wheels

850 IMPLEMATIC

*Rear view of the 850
Implematic*

OPPOSITE:

850 IMPLEMATIC

*The David Brown 850
Implematic*

BROCHURE
880 Implematic Brochure

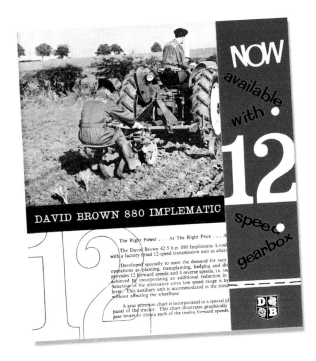

REAR VIEW
Showing the 42 inches of clearance

OPPOSITE
850 High Clear Prototype

DROP BOX
*Rear view showing the drop
box for the rear wheel drive*

The Oliver 500 and 600

THE OLIVER 500 and 600 tractors were David Brown 850 and 950/990 tractors, painted in the Meadow Green and Clover White Oliver livery, with a different radiator cowl.

OPPOSITE:

THE OLIVER 500 AND 600

A David Brown 850 in Oliver livery (The Oliver 600 is a David Brown 950 or 990 in Oliver livery.)

The Oliver Corporation in the USA needed a small to medium sized tractor for the North American market and approached David Brown, as they did not want the expense of manufacturing one themselves. An agreement was reached, which led to the first consignment of David Brown/Olivers leaving Meltham in the early 1960s.

The Oliver 500 (David Brown 850) was sold in both the United States and Canada. The Oliver 600 was sold only in America, the early ones were David Brown 950s. In 1961, the 990 tractor was introduced and so later models were 990s, which meant that they were more powerful, with the horsepower increased from $42\,^1/_2$ to 52.

The arrangement between David Brown and the Oliver Corporation worked well until Ford sacked all their dealers in the USA in favour of a distributor network, which left a lot of unhappy Ford dealers looking for tractors to sell. This was too good an opportunity for David Brown Tractors to miss and meant that the Oliver deal had to be dropped.

BROCHURES
Oliver 500 and 600 sales brochures

David Brown 990 Implematic

OPPOSITE:
990 Implematic

This was one of the most popular tractors built by David Brown, being second only to the Cropmaster. As the 950 was an improved 900, so the 990 was an improved 950.

The 990 was a little longer, it had heavier castings, a stronger back end, a larger clutch and a new engine. It was the first David Brown tractor to have an engine with a cross-flow cylinder head. The horse-power went up to 52, and the gear-box was improved. With having a longer chassis, the air cleaner could now be accommodated just behind the radiator grill, which meant not only cleaner air for the engine, it also tidied up the left side of the tractor where the air cleaner used to be.

As with other tractors in the David Brown range during the 1960s, height control became a feature in early 1963. Later in that year, the two 6-volt batteries situated one each side of the seat, were replaced with one 12-volt battery which was now located under the air cleaner at the front of the radiator. At the same time, the tractor was lengthened by a further two inches and a new fabricated front axle was fitted in place of the old forged one. The two redundant battery boxes each side of the seat were now turned into toolboxes. By the end of 1963, a twelve speed gearbox was offered as an option.

In preparation for the Selectamatic hydraulic system, the 990 underwent further changes in mid 1965 when heavier rear axle castings were used. The toolboxes each side of the seat disappeared, and the seat was mounted on supports from the transmission housing. A new drawbar pickup hitch and mudguards were also fitted.

990 IMPLEMATIC: *Hoeing and weeding simultaneously*

990 BROCHURE
990 sales brochure

David Brown 770 Selectamatic

PRODUCTION OF THE 2D FINISHED IN 1961 but David Brown Tractors were still convinced that there was a market for a small tractor:

PRODUCTION DETAILS

The David Brown 770 Selectamatic

Built from: 1965\1970
No. produced: 12,206

OPPOSITE
770 Selectamatic photographed near Meltham in typical David Brown countryside

The big problem, from a manufacture's point of view, was that it was often cheaper for the farmer to buy a slightly larger second-hand tractor, than one ideally made for the job. Against this background, the 770 Selectamatic was introduced in early 1965. It was powered by a three-cylinder engine of 33 hp, it had a twelve speed gearbox operated by two levers. The 770 was the first tractor in the David Brown range to be fitted with Selectamatic hydraulics, a device that enabled the driver to select, by simply turning a dial, whichever hydraulic function was required, for example, draft control, position control and external services.

This system of hydraulic control was so successful that later in 1965 it was fitted to all the models in the David Brown range. At the same time, all models were restyled and the now famous Hunting Pink with yellow wheels was replaced by Chocolate Brown with Orchid White panels and wheels. In addition to the revamp, the 770 received an engine upgrade. By doing away with the dry cylinder liners and using a bored cylinder block, the power was increased to 36 hp.

REAR VIEW
Rear view of a 770 Selectamatic

DB 770 SELECTAMATIC
The 770 Selectamatic – introduced in spring 1965 (The 770 Selectamatic changed colour in autumn 1965, when all David Brown's changed to Orchid White and Chocolate Brown.)

770 SELECTAMATIC

*The David Brown 770
Selectamatic*

DECALS

The 770 Selectamatic Decals

design features

★ 33 BHP direct injection 3-cylinder diesel engine.
★ 30 pto horsepower.
★ 12 Forward and 4 reverse speeds.
★ Differential lock.
★ Single (Standard) or dual clutch (Livedrive).

★ Selectamatic hydraulics — any one of four hydraulic services selected at the flick of a switch
★ 540 rpm pto unit to British and S.A.E. standards.
★ Category I linkage with mechanical lift latch.

★ Adjustable cushioned seat.
★ Feather-light recirculating ball steering.
★ Key-operated 12 volt starting.
★ Unit construction—with all engines, gearboxes, and hydraulic pumps and valves tested before assembly into the tractor.

BROCHURE

*David Brown 770
Selectamatic slaes brochure*

TWO VIEWS

*Two views of a white and
chocolate Selectamatic*

770 SELECTAMATIC: *with David Brown front-end loader*

David Brown 780, 880 and 990 Selectamatic

From October 1965, the 880 and 990 were restyled and painted in the new colours of Orchid White for the panels and wheels and Chocolate Brown for the engine and transmission.

880 SELECTAMATIC
View of right side

PRODUCTION DETAILS

The David Brown 880 and 990

Built from: 1965\1971

David Brown 780 Selectamatic

Built from: 1967\1971
No. produced: 12,198

At the same time the Selectamatic hydraulic system became a standard feature. As with the 770, by simply turning a dial the operator could select one of three different hydraulic functions: depth control, height control and external services. The engine horsepower ratings were increased to 46 hp on the 880 and 55 hp on the 990. Both tractors had a six forward and two reverse speed gearbox, differential lock and a two speed power take-off as standard equipment. A twelve forward and four reverse speed gearbox was available as alternative equipment, as was a conversion kit to turn these tractors into high clearance tractors and back again, as the situation on the farm demanded. In 1967 the engine of the 990 was changed from a wet liner type to one with a bored block similar to the engine used in the 1200 but with a shorter stroke. This engine is recognised by the oil filler cap located on the timing case and a separate dipstick at the side of the engine. On the older engine the oil filler cap and dipstick were combined.

From 1970, the 990 was available in four-wheel-drive form and full flow filtration of the hydraulic oil became a standard feature on both models.

The 780 used the same 46 hp engine but was available only as a live drive with a dual clutch and a six speed transmission; twelve speed operated by two levers was an option. In 1969 a narrow version was introduced.

VIEW FROM THE DRIVER'S SEAT
of an 880 Selectamatic

RIGHT
780 Narrow

A 990 Selectamatic equipped with Lambourne cab and modified loader from an Implematic tractor

990: *A 990 Selectamatic complete with Sta-dri cab*

David Brown 1200 Selectamatic

WHEN INTRODUCED in 1967 this tractor was David Brown's flagship model of 67 hp, designed to compete with the Ford 5000, the Massey Ferguson 175 and the International Harvester 634.

FOUR WHEEL DRIVE
A fine example of a relatively rare David Brown 1200 four wheel drive

PRODUCTION DETAILS

The David Brown 1200 Selectamatic

Built from: 1967\1971
No. produced: 18,990

Turned out in the now familiar Orchid White and Chocolate Brown colour scheme with 12x38 rear tyres and 7.50x16 fronts, it was a most impressive looking tractor. This was the first David Brown model to have a separate hand clutch to operate the power take-off.

Another first was that the hydraulic pump, now located in front of the engine, was crank shaft driven. The three-point linkage was category two only and, as the name implies, the Selectamatic hydraulic system was used. Early Selectamatic tractors had an exhaust brake as standard, as the brakes were the same as the 990.

When the brake size was increased, the exhaust brake became an optional extra.

In 1968 the engine was uprated to 72 hp, and a four-wheel-drive model became available in 1970. Standard specification included a six speed gearbox, a super-deluxe suspension seat, multi-speed power take-off and all the usual features such as lights, differential lock, horn, tractormeter, and so on.

Optional extras included twelve forward and four reverse speed gearbox, power steering, exhaust brake and hydraulic valves.

1200 SELECTAMATIC

The 1200 Selectamatic – introduced in 1967

STA-DRI CAB

1200 Selectamatic with Sta-Dri cab

David Brown 1212 Hydra Shift

FITTED WITH the same four-cylinder 72 hp engine as the 1200 Selectamatic, the 1212 was the first tractor to be fitted with the famous Hydra Shift semi-automatic transmission.

PRODUCTION DETAILS

The David Brown 1212 Hydra Shift

Built from: 1971

This provided the driver with clutchless gear changes between four ratios in any preselected gear: creep, field, road and reverse. The clutchless changes were made by simply moving a small lever on the dash. Even when used in the field, changes could be made smoothly and under full load. When used on the road, full engine braking was maintained in each ratio.

This transmission won several awards for the David Brown Tractor company. In1964, it earned the Queen's Award to Industry for Technological Achievement. This was the first time an agricultural tractor manufacturer had gained such an honour. In 1976 the Hydra Shift transmission was granted a Design Council Award, another rare achievement.

The 1212 was the first David Brown tractor to be fitted with hydrostatic power steering as standard. Until this time hydrostatic power steering had been available only as an option on some models. Selective top-link sensing was another standard feature.

1212 HYDRA SHIFT
from the driver's seat (The Hydra Shift lever can just be seen under the steering wheel, top right.)

OPPOSITE
1212 Hydra Shift with full set of front end weights

David Brown 885, 990, 995, 996 and 1210 Synchromesh

THE 885 REPLACED the 780 and the 880 by using the best features of each. Like its predecessors it was also available in a narrow form.

885
An 885, with safety frame removed, ploughing some heavy land

PRODUCTION DETAILS

The David Brown 885, 990, 995, 996, 1210 Synchromesh
Built from: 1971

The main features included: a 48 DIN hp engine giving 44 PTO hp; live multi-speed power take-off, providing 540 and 1000 rpm; twelve forward and four reverse gears, with synchromesh on eight of the forward gears; Selectamatic hydraulics and dual category linkage. Standard tyre sizes were 6.00x16 on the front and 11x28 on the rear.

The 990 Synchromesh now had 58 DIN hp, giving 53 PTO hp. Other features included all of those of the 885. Standard tyre sizes were 600x19 on the front and 11x36 on the rear.

The 995 and the 996 Synchromesh tractors were more powerful versions of the 990: both had 64 DIN hp engines, providing 59 PTO hp; and both had all the features of the 885. The 996 had a separate hand clutch to operate the PTO, in the place of the dual clutch on its smaller brothers. Standard tyre sizes were 7.50x16 on the front and 12x36 on the rear.

The 1210 Synchromesh replaced the 1200. It had a 72 DIN hp engine giving 65 PTO hp and incorporated most of the features of the 885, with the exception of the dual category linkage, the 1210 was category two only. Standard tyre sizes were 7.50x16 on the front and 12x38 on the rear.

1210

(This is the same tractor as the 1212 but with a synchromesh gearbox instead of the Hydra Shift.)

1210 - *rear view*

885 NARROW
The 885 narrow version

1210
*A 1210 fitted with a Turner
Hydra Mower 15 hedge cutter*

SAFETY CAB
An 885 with full David Brown safety cab

IMPROVED ACCESS: *An 885 fitted with full David Brown cab with improved access*

ABOVE: *An 885 Selectamatic bringing in a load of bales*

990 *A David Brown 990 Selectamatic complete with David Brown loader still earning it's keep*

SAFETY FRAME: *An 885 fitted with safety frame and Vicon fertilizer spreader*

A 990 fitted with full cab

DAVID BROWN EXPERIMENTAL TRACTOR
A power red and white 995 with full cab, front end weights and rear wheel weights (This tractor was a David Brown experimental tractor.)

996 - with safety frame (The difference between the 995 and the 996 is that the 996 has an independent hand clutch to operate the PTO.)

995
A chocolate and white 995 with safety frame

995 - A 995 with trailer

David Brown 1410 and 1412

David Brown 1412 with New Holland Forager and Weeks Twin 7 trailer

The 1410 and the 1412 were David Brown's first production tractors with turbo charged engines, a four-cylinder unit producing 91 DIN hp which gave 81 PTO hp.

Other firsts for this tractor were the dry air cleaner for the engine; oil immersed disc brakes; telescopic lower lift arms; and new steering geometry with the hydrostatic steering ram situated crossways, just behind the front axle. The 1410 Synchromesh, as the name implies, incorporated the twelve forward and four reverse speed gearbox, plus the other features of the 1210. In 1976 a four-wheel-drive version of the 1410 was introduced.

The 1412 had all the above features, but in the place of the twelve speed gearbox the hydroshift gearbox was used which, as mentioned earlier, gave four clutchless gear changes in four different ranges. The standard tyre size for both tractors was 7.50x16 on the front and 14x34 on the rear.

1412
David Brown 1412

The Silver Jubilee Tractor

THE YEAR 1977 was not only the year of Queen Elizabeth II's Silver Jubilee, it was also the year that the 500,000th David Brown tractor was going to be produced.

THE SILVER JUBILEE TRACTOR
The Silver Jubilee tractor with gold cab for the Golden Jubilee celebrations, 2002

It was therefore decided that the tractor would be auctioned at the Smithfield Show that year and the proceeds would go to Her Majesty the Queen's Jubilee Appeal.

The tractor would be a 1412, which unfortunately was not going to emerge from the production line until November. There was no alternative but to use another identical 1412 tractor for the publicity photographs.

The Jubilee Tractor was to have a silver cab with a Royal Purple roof, the publicity tractor would therefore have to receive the same treatment.

At the Smithfield Show, the tractor had its own stand prominently positioned on the main David Brown stand. All the publicity, posters and leaflets were out announcing the 500,000th David Brown tractor. Only the sharp-eyed show-goers noticed that the tractor on the stand was on Goodyear tyres but the one on the posters was on Klebber tyres.

At the end of the show, the tractor was auctioned and £16,000 was raised. The Jubilee Tractor was bought by Mr Peter Emmett, who used the tractor on his farms for several years. Having paid £16,000 for a tractor with a list price of £10,000, he is reported to have said, "It will have to earn its keep."

The next public sighting came when J Gibbs of Bedfont advertised the David Brown Jubilee Tractor for sale by tender. The successful bidder, on this occasion, was Mr Francis Brown of Stembridge, Somerset. Unfortunately, shortly after acquiring the tractor, Mr Brown's health declined and the tractor found its way, via an indirect route, to Mr Eddie Hocking's workshop in Penzance, Cornwall, where it was partly restored. The tractor changed hands again in 1998, this time Mr Patrick Palmer of Bower Hinton Farm, Martock, was the purchaser. He engaged Eddie Hocking to finish the restoration. In June 1999, Mr Jim Espin became the proud owner of this unique tractor.

For the year of the Queen's Golden Jubilee 2002, the cab was painted gold where previously it had been silver.

SILVER CAB

The 1412 Silver Jubilee Tractor with its silver cab and purple roof

David Brown 90 Series

BY THE MID to late 1970s, the David Brown tractor range was beginning to look a little bit dated. A new range of tractors with a new image, which would incorporate both David Brown and Case tractors, was needed.

HAY TEDDING
A 1690 hay tedding (This is usually a job for a smaller tractor but, when the sun shines, whatever is available has to be used)

1690 Four-wheel Drive (The 90 series was the last to carry the David Brown name.)

O n 26 September 1979, David Brown Case announced the new 90 series to its dealers, who had been taken to a high class hotel in Monte Carlo for the grand occasion. Tractors were displayed on several floors of the hotel.

The complete range consisted of five David Brown tractors, ranging from 48 hp to 105 hp, built at Meltham and five Case tractors, with horsepower ranging from 120 to 273, built in America. All tractors shared the same colour scheme of white panel work with the engine, transmission and wheels in Case Power Red (orange).

The Meltham built models all had new cabs designed by Sekura in conjunction with David Brown and built by Sekura. For the late 1970s, the cabs were quite luxurious with two wide access doors that were hinged at the rear, two good steps each side, a flat deck cab, luxury seat, a heater and windows that gave good all-round vision. The view forward was second to none.

The smallest model in the 90 series was the 1190, with 48 hp, twelve speed Synchromesh gearbox, hydrostatic power steering and multi-speed power take-off.

Next came the 1290 with 58 hp and then the 1390 with 67 hp (essentially the same tractor but with 9 more horsepower for farmers who needed a little more power). Both tractors had the same basic specification as the 1190. Optional equipment on these two tractors included four wheel drive and a high clearance conversion kit which raised the tractor 6 $\frac{1}{4}$ inches. This kit consisted of a special pair of front axle beams that raised the front of the tractor, the rear end was raised by rotating the final drive units. This could be done on the farm, when the tractor was required for high clearance work, and converted back again for general duties. Hydra Shift transmission became available for the 1390 in 1983.

The 1490 had the largest four-cylinder engine, which was turbo charged to give 83 hp. The claim to fame for this engine was that it was designed for turbo charging with a large diameter crank shaft and bigger oil and coolant capacities. This tractor was equipped with the famous David Brown twelve speed synchromesh gearbox and multi-speed power take-off as standard. Optional was the Hydra Shift transmission on two-wheel-drive models, four wheel drive was another option, along with a high clearance kit.

The 1690 was the big boy in the Meltham stable: 105 hp, six-cylinder turbo charged engine. This model was available only in four-wheel-drive form. It featured lower link depth sensing for more accurate control of large mounted implements, twelve forward and four reverse speed synchromesh transmission, and a fully independent power take-off providing both 540 and 1000 rpm.

The original four-wheel-drive units for all 90 series tractors were of David Brown manufacture. In 1982 Carraro four-wheel-drive axles could be fitted. These had the disadvantage of increasing the turning circle. Also in 1982 a new bonnet, which was hinged from the front and so gave much better access to the engine, was introduced. The headlights were mounted higher in the new bonnet to give a better light spread for night-time driving.

SEKURA SAFETY CAB
1290 with the then new Sekura safety cab

AN EARLY 1490
with headlights in the bottom of the grill

A 1490 TRACTOR

The David Brown 1190 was more suited to light work as this picture with a Vicon Acrobat hay rake illustrates.

1290

The David Brown 1290 seen on typical livestock duties

1390

Towing a trailed crop sprayer

1690

David Brown 1690 four wheel drive and Dowdeswell plough

1490

David Brown 1490 baling hay with a New Holland Super Hayliner and Browns Juggler accumulator

The David Brown Case 94 Series

THE 94 SERIES, which was introduced in 1984, included: the 1194 with 48 hp; the 1294, 61 hp; the 1394, 72 hp; the 1494, 83 hp; the 1594, 95 hp; and, finally, the 1694 with 108 hp.

DAVID BROWN CASE 1194

All the engine blocks on the four larger 94 series have the cylinders cast into them. The bore and stroke dimensions, which are the same for all four models, are now given in metric, they are 100x114.3 mm. As many parts as possible were interchangeable among the four models, in order to reduce the amount of stock needed by the individual dealers.

The 1394, 1494 and 1694 were turbo charged. The 1194, 1294 and 1594 were naturally aspirated. The 1194, 1294, 1394, 1494 and the 1594 had a twelve speed synchromesh gearbox as standard. A creep speed option was available for the 1394 and the 1494 – this gave three more gears that were slower than the standard first gear. The 1694 had the hydroshift semi-automatic gearbox and four wheel drive as standard features. These features were also available as options on the smaller models, apart from the 1194 which was two wheel drive only.

All six models had two speed power take-off, the lower range of which was 540 rpm and the higher range, 1000 rpm. The PTO was controlled by an independent hand clutch, with what was described as low-lever effort, so that power could be easily controlled independently of the forward motion of the tractor.

Hydraulic power comes from an engine driven pump. The 1694 and all four-wheel-drive models had tandem pumps. These pumps increased the oil flow to enable hydraulic motors or multiple functions to be operated simultaneously. The 1194, 1294, 1394 and the 1494 have top link depth sensing whereas the 1594 and the 1694 have lower link sensing.

The 94 series were the first Meltham produced tractors not to carry the David Brown name – just the name of Case on the bonnet, with the model number on the mudguard. When the 94s were introduced, a new colour scheme was used: white panel work, black engine and transmission, and orange wheels. A high clearance conversion kit was available for the 1294, 1394 and the 1494.

In 1985, Tenneco acquired the International Harvester Company. This meant that there were two tractor plants within a few miles of each other, under the same ownership and competing for the same market. The result was rationalisation – the 1194 and the 1294 were discontinued. Another colour change came in 1986 when the colour scheme was standardised with International Harvester: red panel work, black engine and transmission, black and silver wheels. Tractor production finally came to an end at Meltham Mills in 1988 but not before a number of 94 series commemorative edition tractors had been produced.

AMERICAN ADVERTISING BROCHURE

An American advertising brochure for the David Brown Case 94 Series

1394 COMMEMORATIVE EDITION
– to commemorate the end of tractor production at Meltham Mills

SANDS SPRAYER
Sands sprayer built on a David Brown 1494 skid unit

1594: *With Bomford Highwayman hedgecutter*

1294: *1294 four wheel drive export model (note continental hitch and assiter rams)*

1294
export model

1494
Four wheel drive

1494 *Two wheel drive*

1594

Four wheel drive with four furrow reversible plough

1594

*The last tractor built at
Meltham outside Durker
Roods Hotel the one time
home of Sir David Brown*

Implements

DAVID BROWN Tractors have made implements since they made the first ploughs and cultivators for the Ferguson Brown tractor in 1936.

SINGLE FURROW PLOUGH
David Brown single furrow plough

DOUBLE FURROW PLOUGH
David Brown double furrow plough

SINGLE FURROW ONE-WAY
PLOUGH
David Brown single furrow one-way plough

When the VAK I tractor was introduced the range was extended, and by the time the Cropmaster came in, the range included a one-way plough, several ordinary ploughs, disc ploughs, disc harrows, mowers and a potato spinner.

CULTIVATOR
Two different types of David Brown cultivator

In 1955 David Brown acquired the firm of Harrison, McGregor & Guest Ltd of Leigh in Lancashire, and with this company came the Albion range of farm machinery. Although much of the machinery was out of date, David Brown now had the space and opportunity to make implements and machines to suit their range of tractors. Very soon the horse-drawn tackle was discontinued. A modern range, which included some of the old Albion range, was manufactured. Under the David Brown Albion banner, ploughs from one to six furrows, disc ploughs, cultivators, ridgers, a rotary tiller, a subsoiler/mole drainer, manure spreaders, seed drills (12, 14 or 15 rows), mowers and swath turners were all made. Several larger machines were also produced. The baler with its unusual crossways mounted bale chamber, which was only eight feet three inches by six feet three inches in length (without its drawbar), could do a big job but be stored in a small space.

The Hurricane forage harvester was available as an inline or offset machine. It was PTO driven and could cut, chop and load the crop in one operation. The Albion binder was very popular as it was available in land wheel drive or power take-off drive and could cut widths ranging from four feet six inches to ten feet. Some 67,000 of these binders were produced before they were phased out in 1958.

The David Brown Albion combine, which was introduced at the 1955 Smithfield Show, was made and sold by Arvika Verken in Sweden as their BT 45. It was imported and assembled by David Brown at Leigh. A five foot cut trailed machine, which was PTO driven, was available as a bagger or a tanker.

During the 1960s, the Leigh factory output was gradually changed from machinery to tractor components. Many of the implements were discontinued; some were taken on by other manufactures. The process continued as tractor production increased to meet the demand from overseas. At one time four out of every five tractors built were exported. By the mid 1970s the only machines produced were front end loaders and rear mounted diggers.

MID-MOUNTED MOWER
David Brown mid-mounted mower – hydraulically driven, and raised and lowered

EARTH SCOOP
David Brown Earth Scoop

POTATO SPINNER
David Brown Potato Spinner

LOADER
David Brown loader for the Implematic range

MANURE SPREADER
The David Brown farmyard manure spreader

THE DAVID BROWN ALBION

ROTARY TILLER

(The tiller was made in three widths: 90 inches, which cost £86 15s 0d; 106 inches, which cost £95; and 118 inches, which cost £99 10s 0d.)

BALER

The David Brown baler with its crossways bale chamber

131

DAVID BROWN ALBION PICK-UP BALER
The late Chris Grant, a David Brown enthusiast with his treasured 880 Implematic and David Brown Albion pickup baler. Chris sadly died at a relatively young age after a long illness

BALER
The David Brown baler with its bale chamber extension folded for transport

ALBION PICK-UP BALER
The front view of the David Brown Albion Pick-up baler

BALER DETAIL
David Brown Albion baler clearly showing the side discharge bale chamber

INDUSTRIAL TRACTOR

*David Brown Industrial tractor –
equipped with David Brown front
end power loader and David
Brown digger on the rear*

FRONT WHEEL BRAKE
– fitted to Industrial tractors

ABOVE AND OPPOSITE: *David Brown Tractor - in Case Power Red and Orchid White – fitted with David Brown front end loader and rear mounted digger*

EWX 493

POTATO SPINNER
A potato spinner mounted on a VAK 1

Steerage Hoe
An early steerage hoe mounted on a VAK 1
equipped with row crop steel wheels – operated
by two land-girls in period dress

HARVEST SCENE
– David Brown tanker combine and David
Brown baler working together

HARVEST SCENE
– 990 Implematic with David Brown bagger combine

141

BALING EMPTY FEED BAGS
– *an unusual chore for any baler!*

DAVID BROWN 950 IMPLEMATIC
*David Brown 950 Implematic hard at work –
demonstrating a David Brown Albion
Hurricane forage harvester*

TURBO TILLER

The Turbo Tiller stubble cultivating behind a 950 Implematic while the combine is busy in the background

ROTARY CULTIVATOR

Rotary cultivator operating behind a 990 Implematic (Note that there are four rotors, each driven by its own Radicon gearbox.)

THE LEIGH FACTORY

Three production lines at the Leigh factory – combine (far left); balers (centre); and FYM spreaders (right)

MANURE SPREADER
Farmyard manure spreader doing what it does best

Equipment associated with David Brown

JONES BALER
Jones baler powered by a Cropmaster engine mounted underneath

BONSER FORK TRUCK
A Bonser fork truck powered by a David Brown three-cylinder engine

SACK LIFTER

RIGHT

Cropmaster equipped with Mill loader

HORNDRAULIC LOADER

Super Cropmaster equipped with Horndraulic loader

RIGHT

Cropmaster with Sky High loader

OPPERMANN WHEEL STRAKES

STA-DRI

950 with Sta-Dri tractor cab

LOAD CAR
The load car used for dynamometer testing

REAR AXLE AND HYDRAULICS
The wheel had turned full circle (David Brown started tractor production by making the Ferguson Brown. By the mid 1990s, David Brown Engineering were making the rear axle and hydraulics for the American two-wheel drive Massey Fergusons.)